How To Get AI OUT of Your Classroom

ARVIN VOHRA & THE TURING POLICE

ISBN: 978-0-9992711-8-6 (Paperback)

ISBN: 978-0-9992711-5-5 (Ebook)

Library of Congress Control Number: 2026937887

Written by Arvin Vohra and The Turing Police

Book Cover and Typesetting by Lidia Pucetti

www.TheTuringPolice.com

TABLE OF CONTENTS

INTRODUCTION

Not long ago, I was ready to quit teaching.

I've been called to teaching since I was in elementary school, helping my classmates with math. I started formal volunteer tutoring when I was 15 and started doing paid tutoring during college. After college, when almost all of my friends went into investment banking or management consulting, I decided to start a tutoring business. Soon, I started doing group classes, and a few years after that, I created the current group classes that I teach. Like many people who are obsessively passionate about what they do, I named my classes after myself, calling them Vohra Method classes. Years of work and obsession have gone into every part of the curriculum. It's evolved from a highly experimental in-person class to a continually evolving online program.

To say that I'm passionate about education is a massive understatement. But not long ago, I was thinking of quitting teaching. Instead of spending nights thinking about ways to improve my program, I was spending nights looking at job postings.

You might be thinking of quitting teaching for the same reason. AI had made my job impossible and miserable. It had turned this program that I had spent the majority of my adult life working on utterly ineffective. My once engaged, energized students were sullenly checked out, but "somehow" getting all the right answers with ease...and writing them in flowery prose with perfect grammar. I suspected that those answers were AI-generated, but I had no way to prove it. AI checkers were useless. Students had access to the same AI checkers, and could just adjust the wording until the AI checker registered their answers as human-generated.

My classes are online, so if a student is in class, he has all the AI in the universe right there on his computer. I had no way to physically block AI and no way to prove AI use. I only had my suspicions and a class full of checked-out, zoned-out teenagers.

I missed the "bad" parts of pre-AI teaching. I missed the one-word answers to complex reading questions, students griping about a problem being too hard, and students just answering "idk." Now, no student was complaining or getting stuck. They were answering in long, professional, grammatically flawless, unprovably AI-generated sentences.

My biggest strength as an educator is curriculum design. I've built both traditional and nonlinear curricula for both in-person and online classes. I've built curricula for my own programs and for other educational institutions. I tend to believe that with a strong enough curriculum, even an average teacher can produce excellence, but that with a bad curriculum, the most charismatic teacher on earth can't.

But my decades of experience in curriculum design meant nothing right now. AI was turning my flagship curricula into junk.

I wasn't limited by laws or policy. I could change whatever I wanted. It was my company, and I had completely free rein. And even with all that, I was getting nowhere.

I was trying one thing after another, and nothing was working. At first, I used GPTZero, a popular AI detection tool, to try to detect AI use. Students easily bypassed it. They just put their own answers into GPTZero and messed around with the wording until the AI detector believed it was human-generated.

I saw answers that I strongly suspected were AI-generated. I saw answers that looked like someone had taken an AI answer and replaced words with strange, inexact synonyms. I was suspicious, but nothing was provable. For a while, I tried treating suspicion of AI use as proof of AI use, but students almost instantly adapted to that. I even built my own AI detector, which didn't fare much better.

Not only are my classes online, but they are based on challenge problems. The problems in my programs take around 10 to 60 minutes to solve. That creates a huge temptation to use AI.

I could tell that more AI use was happening. Problems that for decades had taken students half an hour to solve were suddenly getting solved in 2 minutes. But again, I rarely had any proof. Also, because my programs attract extremely smart and ambitious students, there was always a possibility that any one student had actually figured out the questions that quickly without AI. There was no way every student had done the questions that quickly without AI, but I had no way to tell who was doing what.

I was also seeing disturbing changes in my students. More and more students seemed completely and utterly checked out. Historically, students were engaged during class. Sometimes they would good-naturedly protest the difficulty of a problem. Sometimes they would bad-naturedly protest the difficulty of a question. But now, many of my students seemed barely conscious.

But there was something else I found much more troubling. I was seeing a kind of bald-faced lying I had never seen before in students. While the AI checkers were mostly useless, sometimes they did correctly catch a student. Until now, when I caught students doing something they weren't supposed to do in class, they'd sheepishly laugh and admit it. But now, students caught using AI would just brazenly, unapologetically lie about it. They would maintain the lie in the face of repeated questioning and in the face of irrefutable evidence.

I had never seen that kind of lying in over 20 years of working with students. What scared me was that I had seen that kind of lying somewhere else. It looked exactly like the kind of lying that comes from alcoholics and drug addicts. If you've ever worked with those struggling with addiction, you probably know what I'm talking about: the persistent, brazen, confident lying, no matter how much evidence is against them. As this book discusses later, AI use has several disturbing parallels to addiction.

By the way, my students are "good" students. Most have straight A's at school and take all honors and AP classes. They have parents who care deeply about education. In fact, the students care about education too. They're voluntarily taking a hard course outside of school.

Even with complete freedom to change the curriculum however I wanted, even with motivated students and parents, I couldn't fix my program. I was starting to suspect that running challenging online classes in the AI era was simply impossible.

As the existence of this book suggests, I did manage to figure a few things out. Instead of trying to fix my curriculum, I built a new one from scratch. It was built from the ground up to be AI-proof and AI-resistant. I threw out many of my cherished, old methods. They're obsolete. They're never going to work again. Many were innovations I was particularly proud of. Some were innovations I'd worked on for over a decade. But I had to let them go.

I created new methods. I relied on my past experiences working with students, but I also drew heavily on my experiences working with those struggling with chemical, digital, and process addiction. (Process addictions include gambling addictions and the like.)

Through this process, I learned more than I'd ever imagined about how to create AI-proof and AI-resistant education. It may be of use to you, even if your classroom looks nothing like mine.

In my first class with the new curriculum, I remembered why I loved education. I'd forgotten what it felt like to have a good class. I'd become so habituated to classes becoming awful, tedious, losing battles to AI that a good class felt unfamiliar. The students were energized the way they had been before AI ruined education.

I have much more work to do and many more things to learn about AI-proof education. But I know this: it can be done. I've done it, and I'm working to do it better. I believe that you can get AI out of your classroom as well, and I hope that my experiences and insights will help you.

AI-fueled fraud isn't just a nuisance. Its damage goes past making individual teachers feel powerless. The moral issues here go far past the fact that cheaters are getting unfairly rewarded and that the real accomplishments of students with integrity are insufficiently recognized. It represents a situation in which people are misrepresenting their skill level to universities, and then continuing to misrepresent their skill in ways that put individuals and societies in danger.

Consider the education of medical doctors. There's no question that doctors learn the bulk of their medical knowledge during their residencies, where AI use may be less constant. But the skills they develop before residency contribute in vital ways. The problem-solving skills they develop in organic chemistry and geometry classes contribute, perhaps indirectly, to their medical problem-solving skills. Their analytical skills, understanding of nuance, and development of empathy for those with different experiences are developed through the study of literature. Their communication skills are developed through writing. Their understanding of cultural shifts, relevant to medical outreach and compliance, is developed through a study of history. A doctor who hasn't studied any of these areas, but rather only pretended to, is going to be worse than he could have been. That doctor has also taken the place of someone who actually did the work and developed the skills, who could be a much better doctor.

The AI-fraudster doctor has also failed to develop the most basic level of ethics and integrity. His entire education has been a lie; the skill he has developed above all others is lying. Would you want an unethical, dishonest, uneducated doctor treating you?

And of course, that charlatan's hands-on knowledge is going to be damaged as well. When he presents at Grand Rounds, is his work and analysis going to be his own? Is it even going to be based on reality? Or will he use AI-generated fakery to lie to and damage the medical knowledge of others?

Would you want a doctor who lacks critical thinking skills, creative thinking skills, or any thinking ability at all? Would that physician have the critical skepticism needed to evaluate pharmaceutical marketing, or the analytical ability to question dubious drug trials? Once he becomes a doctor, will he suddenly become able to think, or will he keep relying on AI to diagnose, treat, and communicate with patients?

And this isn't limited to doctors. What about policymakers who never wrote a single history paper or political science paper, and never read any of the assigned readings? What avoidable pitfalls will they miss? What about investment bankers who cheated through every economics assignment? What about journalists who never read a book or wrote a paper on their own? Do we expect them to shed an investigative light on the incompetence of the AI-dependent economists, doctors, and policymakers? What kind of policy and economic disasters is AI-fueled fraud bringing?

Educators have a moral imperative to stop, prevent, and punish AI-fueled cheating. If you're a teacher who is ready to stop AI from destroying education, this book may help you get AI out of your classroom. But, like me, you may need to let go of your most cherished and long-lived methods. As educators, we help students learn to question their assumptions; to get AI out of our classrooms, we will need to question our own assumptions. AI has completely altered the world; we will have to completely alter our methods. We will have to rethink our methods and step far, far outside of our comfort zones.

Put more bluntly: in most schools, students are learning absolutely nothing. Education is facing an existential crisis. Thus, there's no educational method that's sacrosanct. We must be willing to change anything and everything. We must be willing to adapt faster than Silicon Valley does.

Silicon Valley's ethos is to move fast and break things. As educators, we must move faster to fix things. AI has rendered current educational methods utterly ineffective. We're going to have to move fast, innovate relentlessly, and be willing to throw out any obsolete assumption if we're going to nullify AI's ability to destroy education.

If you're an administrator who is losing teachers or hemorrhaging students because AI has made your programs ineffective and miserable, the tools, insights, methods, and approaches described in this guide may help you turn things around. If you work in admissions, these techniques will help you admit the best and brightest students with integrity, not the most fraudulent liars. And if you're a teacher, this book will make it possible to bring real education back into your classroom.

If you're ready to set aside all assumptions and preconceptions about education and curriculum design, to build real intelligence in the era of artificial intelligence, this guide can help.

1. CURIOSITY AND SELF-IMPROVEMENT

It can be easy to see students as utterly indifferent to education itself. “A” students are seen as motivated by grades, and underperforming students are seen as simply unmotivated. When educators see students fixated on grades and eager to forget anything they learned, it can be tempting to see students as utterly incurious and indifferent to any kind of self-improvement.

But evidence outside of the classroom doesn’t back that up. Go to any high school weight room, and you’ll see just how driven teenagers are to improve themselves. And the fires of raw curiosity are still burning in their minds…outside of the classroom. The same students who are utterly indifferent to meanings hidden in literature spend hours looking up, thinking about, and discussing the meanings of the lyrics of their favorite songs.

That curiosity is hardly limited to lyrics. Students fall into rabbit holes on YouTube and Wikipedia, guided by pure curiosity. Advertisers know this: what is clickbait other than an attempt to latch onto that blazing curiosity? In fact, once students finish using AI to complete hated assignments, they use that same AI to explore, guided by pure curiosity.

Everywhere, teenagers are driven by curiosity and self-improvement. In fact, the only places in which that curiosity and spirit of self-improvement are absent are in the very classrooms ostensibly designed to foster those drives.

Today, as AI-using students mentally check out of the

classroom, every educator should ask whether AI is causing students to devalue education, or simply enabling them to check out of tasks they never valued. And we need to ask where we've failed. Have we failed to "sell" the value to students of what we're teaching, or have we failed to ensure that what we're teaching has any value to students in the first place?

This guide will show you how to forcibly prevent AI use. It will also show you novel ways to detect AI use and punish it after the fact. But nothing you can do will have as much value as ensuring that your classroom contributes to each student's self-improvement or inspires their curiosity, and then communicating that value to students.

Despite portrayals to the contrary, teenagers aren't self-destructive idiots. When they see the value in something, they'll put effort into it. Teenagers know that while AI can give answers, it cannot build their intelligence. When students see education as a way to build intelligence, they avoid AI. When they see it as a series of pointless tasks, they use AI.

Obviously, that's a bit of an oversimplification. Students who believe that Ivy League admissions means wealth and anything else leads to poverty will use AI to cheat, especially if they see everyone else doing it. That doesn't mean they don't value intelligence; it only means that they fear poverty more than they value intelligence.

Thus, before you work on ways to block AI use, detect AI use, or punish AI use, you should turn your attention to your curriculum. As educators in the AI era, we must make sure we're not missing the log in our own eyes while finding the motes in our students' eyes.

Is your curriculum designed, as well as reasonably possible, to build permanent intelligence, skill, meaningful knowledge, and ability? Intelligence, here, should be defined broadly enough to include social intelligence and emotional intelligence. A rigorous study of literature develops human intelligence just as well as a rigorous study of math, for example. But busywork develops no intelligence or ability, unless mindless obedience counts as an ability.

The curricula used in my classes go through weekly updates and yearly overhauls. There is no curriculum in the world that cannot be improved. As you gain skill and experience as an educator, your curriculum should correspondingly improve.

Once you've made your curriculum as strong as possible, you then need to clearly, simply, and directly communicate to your students the ways in which what they're learning from you will enhance their skills and intelligence. English teachers advise students to show, not tell. But when it comes to explaining the benefits of your class to your students, show, don't tell won't work in the AI era. You need to tell, tell, tell, show, tell, tell, tell, and then tell a few more times.

Phillips Academy Andover and Phillips Exeter Academy famously use the motto, "Non Sibi," which means not for oneself. But the reality of education is this: the vast majority of students put effort into what benefits themselves. If they see that educational tasks build their skills, intelligence, and abilities, they'll do those tasks; if not, they'll outsource the busywork to AI. That's literally what computers are for: doing tedious tasks that humans don't want to bother with.

The tiny fraction that is pursuing education in the service of others will follow the exact same decision-making process. If they see educational tasks as building their skills in ways that allow them to serve, they'll do the tasks. If not, they'll outsource them to AI.

The biggest mistake you can make here is attempting to tie excellence to future income. The relationship is, at best, tenuous. Does developing empathy through reading literature make you more or less likely to be rich? Do most rich people ever use calculus? The idea that high school academics have value only if it directly contributes to financial gain is preposterous. If you build your argument around finances, you'll lose.

Instead, recognize that intrinsic motivation is always stronger than extrinsic. Recognize that every student would rather be strong than weak and rather be smart than stupid. Every student would want more social intelligence rather than less, more artistic ability rather than less, more mathematical intelligence rather than less. No one prefers to be stupid.

Extrinsic motivation has never worked particularly well. Statements like, "You need to know this because it will be on the exam," are usually understood as, "You should forget it after the exam. It has no real value to you." Statements like, "You might need this for your job," are not merely false statements that erode trust. They also suggest that the knowledge has only transactional value and no intrinsic value. At most, extrinsic motivation leads to short-term, resentful memorization.

But in the AI era, extrinsic motivation doesn't work at all. Any extrinsic reward that some knowledge will supposedly bring only lets students know that they need to be able to use AI to access that knowledge. Decades ago, teachers would argue that mental math was necessary because you might not always have a calculator with you. But today, arguing that you might not always have a smartphone with you is preposterous. Most people carry smartphones either in their pockets or in their hands at all times. Thus, any external reward that any knowledge will supposedly bring will be translated as an utterly needless reminder to keep your smartphone close.

The only kind of motivation, therefore, that has any hope of working is intrinsic motivation. Students need to know how the course makes them smarter, how it improves their minds. The people who fill gyms around the world aren't there because strength might make it easier to carry groceries; they are there to make their bodies better. Let your students know how your class will make their minds better.

Excellence has never needed an excuse to exist. The kids lifting weights in the high school gym aren't all training for a sport; plenty are just building fitness for its own sake. Students don't need some secondary motivation to develop more intelligence; intelligence has its own intrinsic value. As long as your program is making students smarter, and they know it, they'll have the incentive to do the real work.

In high school gyms, you don't see teenagers using forklifts to lift weights. They put the effort in because they believe that it will build their own strength and ability. If students see your classroom that way, they'll be more inclined to do the same.

That's the starting point of making any classroom free of AI. But unfortunately, it's not the ending point. A dedication to excellence doesn't erase the possibility of cheating. Olympic athletes, some of the most excellence-focused people on earth, sometimes cheat. Even there, mechanisms have to be created to screen for steroid use, blood doping, and other forms of cheating. Even Olympians fall prey to addiction; as discussed later, AI is inherently addictive.

Thus, in addition to focusing on the inspirational power of

education that truly embraces excellence, we must explore methods to prevent, detect, and punish AI use.

2. WHY STUDENTS USE AI

In May of 2025, the experience of Stephen Cicirelli, a lecturer at St. Peter's College, was widely reported. He caught and confronted a student who had used AI to write a paper. Then, that same student wrote an apology letter...also using AI.

Stories like this can make it seem as if students have turned into lazy jerks. But there's much more to the story.

AI is addictive. Opioids are addictive because they turn off physical pain. AI is addictive because it turns off a mental pain: the pain of being wrong. The fear of being wrong is nearly universal, and AI erases it. With AI, you can never be wrong. The AI can be wrong, but that's not you being wrong. With AI, you get the benefits when it's right, and none of the embarrassment or responsibility when it's wrong.

Just as heavy opioid users lose the natural ability to manage pain, AI users lose the natural tolerance for being wrong. They cannot bear being wrong and thus cannot risk being wrong. Like any addict, they have lost the ability to make conscious decisions about usage.

You may have witnessed students unhesitatingly, brazenly lying about AI use, even when the evidence is incontrovertible. That type of lying is rare among students, but common among addicts. In fact, lying itself is a litmus test for addiction. If you're lying about a behavior, there's a higher chance you're addicted. Alcoholics frequently lie about how much they drink. Most other adults don't.

Before AI, few A students cheated at all. Fewer cheated and then repeatedly lied about it. Today, the majority of A students cheat with AI and then repeatedly lie about it. Even at Ivy League colleges, students use AI to cheat and then repeatedly lie about it. That level of continuous lying is an indication of addiction.

Not only is AI addictive, but it's a new addiction, one unlike any addiction humanity has ever faced. There is no infrastructure or support system to help students break that addiction. There is no "AI Addicts Anonymous," no AI Addiction counseling or rehabilitation centers. Just as drug addicts often want to be clean but are unable to get clean without help, plenty of students want to break their AI addiction, but can't.

That's not to say that all AI abusers are addicted. Some are responding to other pressures. For example, AI has fueled a "difficulty inflation" in many classes. In classes where nearly all students are using AI, teachers often get a completely inaccurate sense of the appropriate difficulty level for the course.

For example, in a physics class, a teacher might assign a problem set of moderate difficulty. Since all students are using AI, everyone gets every question right. For the next problem set, the teacher increases the difficulty; however, because every student is using AI, every student still gets every question right. In response, the teacher makes the next problem set even harder, with the same result. Subsequent ones get harder and harder, until the difficulty level is unreasonable. Now, the only way for a student to survive the class is to use AI on every assignment!

In other cases, students feel competitive pressure to use AI. They believe that if everyone else is using AI, and they don't, they'll lose out on college admissions. They're like competitive athletes in a sports league in which everyone is using steroids. They feel like they have a choice between cheating using AI and keeping up, or avoiding AI and losing.

Thus, the three biggest drivers of AI use are addiction, necessity, and competitive pressure. Later sections of this guide will consider aspects of each, but the first step for all of them is the same: as much as you can, have students do work in class instead of at home. Choose homework over classwork.

If you have a physical classroom, you can easily prevent AI use by disallowing any computer or phone access. That blocking not only prevents AI use, but it also helps you determine what students can reasonably do in a given length of time. Thus, it eliminates the difficulty inflation that happens when teachers target assignment difficulty to the ability of AI rather than the ability of students.

But even more important than physically preventing AI use by relying on classwork is communicating to your students that they can come to you for help with breaking an AI addiction. At the moment, there is nothing similar to Alcoholics Anonymous for AI use. There are no AI rehab centers. No schools have AI addiction counselors yet. But your institution can help lift willing students out of that addiction.

Students who want to break an AI addiction should know they can come to you, or to another specific person at your institution. Every student should also know specifically how to contact that person.

For example, your school might have a specific point person for help with AI addiction. That person's email address or office hours should be posted prominently around campus. Any student who is ready to step up and free themselves from addiction should have a clear path for what exactly to do. Adolescents dealing with a novel addiction and who are willing to put effort into breaking that addiction should be supported and helped. They shouldn't be spending additional effort trying to figure out who, if anyone, can help them.

As with any addiction, some people will realize they have a problem and ask for help. Make sure that help is easy to access.

Others will need to hit rock bottom before they're ready to change their lives and break their addiction to AI. Later sections will show you how to detect and punish AI use and hasten that journey to rock bottom for students who prefer to continue cheating. Empathy for students who fell into addiction and want to get out of it will, unfortunately, need to be paired with utter mercilessness toward students who attempt to keep cheating. You will stop some AI fraud with kindness and understanding. You will be able to prevent some by using classwork instead of homework. The rest will need to be detected and punished after the fact; those punishments may help the punished students become willing to change, while serving as warnings to other students.

3. HOW AI USE DAMAGES EDUCATION

Learning can be separated into three categories: active learning, passive learning, and practice. Passive learning involves taking in information, such as by reading a textbook or listening to a lecture. Active learning involves figuring out difficult problems and includes everything from a long geometry proof to creating a thesis for an interdisciplinary paper. Practice means repeatedly doing problems after first learning how to do them.

Active learning is the best type of learning. Students learn more from wrestling with tough math problems or nuanced papers than from listening to lectures. Passive learning is the worst type of learning.

One popular saying, apocryphally attributed to Confucius, says, "I hear, and I forget. I see, and I may remember. I do, and I understand." Active learning surpasses all other types of learning.

AI doesn't erase all learning; it turns active learning into passive learning. It takes the best type of learning and turns it into the worst type of learning. Instead of writing a paper or solving a problem, the student reads over the paper or the solution to the problem.

But AI doesn't have the same effect on passive learning. Passive learning is just as passive once AI is done with it. It's no better or worse.

Thus, as a basic first step to making your class resistant to AI, consider shifting all passive learning to homework and all active learning to classwork. Recorded lectures can be watched at home, for example, while problems normally assigned for homework become classwork.

All problems in math and science should be done as classwork. All primary source documents should be read and analyzed during classwork, while textbooks should be read as homework.

Reading and understanding primary sources is challenging and requires students to figure out what the text literally means. A textbook is only challenging if it is poorly written. Thus, reading primary sources should be classified as active learning, while reading textbooks should be classified as passive learning.

During classwork, students should be actively figuring out problems. The teacher should not be handing out answers. With AI, students look up answers after trying a problem for a fraction of a second. Good education should be the opposite: students should learn to stick with individual problems for entire class periods.

4. PREVENTION

Making sure your curriculum builds intelligence and inspiring your students to pursue that intelligence is critical to getting AI out of your classroom. But it would be utterly naive to think that educators can end there. Educators should find ways to grade in ways that don't allow even the possibility of AI use. In addition, educators should have the tools to detect and punish any AI use that does slip through.

Blocking AI: In-Person Classes

Physically blocking AI use in an in-person class is technically simple. It starts with making grades based entirely on in-class assessments. This can include quizzes and tests.

It can also include some types of papers. A 5-6 page paper can be written in class, as long as a full day is set aside for it. That can be done on a weekend, like an SAT, or on a weekday, like final exams. A full day is, realistically, longer than what most students spent on papers even before they used AI to write them.

"Writing Days" require that the student knows the subject well beforehand, but does not have the specific essay question beforehand. For example, if the paper is a literary analysis paper, the student should have read and thought extensively about the book beforehand. A single day is enough time to write a paper, but it's not enough to read a complex work of literature, think about it, and then write a paper.

The topic of the paper should be given at the start of the Writing Day. The topic should be specific. If the topic is general, students will start memorizing generic AI-generated essays about the book. Also, students who are absent and need to make up the paper must be given a completely different topic.

Students should have zero access to AI or to the internet during the Writing Day. They should also have no access to Grammarly or other writing enhancers. The assessment should assess what students know, not what Grammarly knows.

Students who have accommodations should be given more time during the same day. In other words, they should be allowed to write into the evening. They should not be allowed to go home, memorize an AI paper, and regurgitate it the next day.

Accommodations will be a significant potential vulnerability. As Rose Horowitz reported in *The Atlantic* on December 2, 2025, accommodations are heavily abused by wealthy and ambitious parents seeking to give their kids an edge. Those same parents are likely to try to find a way to give their kids AI access. The most likely way will involve finagling a way to have extended unsupervised breaks.

During a Writing Day, security will be essential. It will have to be much higher than during an SAT, since SAT breaks happen after a section is finished. Students cannot see a question, take a break, and then return to the same question during an SAT. But during a Writing Day, there's only one question. Any break can be spent using AI to help with that question.

Thus, during a Writing Day, students must have no cell phone or internet access. They should either write essays on paper (like AP exams) or on computers with no internet or AI access. Schools should also use electronic cell phone detectors capable of detecting a turned-off cell phone. Having students turn in a cell phone is insufficient, as parents will happily buy students multiple cell phones to give them an AI advantage. Cell phone detectors are reasonably cheap and effective.

Before the day starts, all bathrooms and other secluded areas should be searched for cell phones. Students may try to preemptively hide phones and access them during the Writing Day. If possible, you may want to arrange for a system by which students can be escorted to and from the bathroom.

If that sounds like a lot, remember this: there are illegal drugs in prisons. Even complete lockdowns, searches, and draconian punishments are not enough to completely block drug trafficking. Unless your institution is more secure than Riker's Island, you're vulnerable to AI use.

By grading in-class assessments only, you can make grades relatively impervious to AI use. But you might feel nervous about not grading homework assignments. Don't be nervous. Many countries, including the UK, don't make homework grades part of the course grade. In many countries, homework is seen as preparation for the in-class assessment, not as its own assessment. It's similar to how athletes use workouts. Athletes exercise in order to perform better during the contest; they get no points for the exercise itself.

What Writing Days Will Reveal

After your first Writing Day, you may think that something went wrong. You may wonder if writing papers in school is so stressful to some students that it is causing neurological damage. You'll see writing so bad, so grammatically bizarre, so incoherent that you may think there is some fundamental flaw with the concept of a Writing Day. You may think there is no other explanation for the ungrammatical, incoherent nonsense that straight-A students are producing.

There is one: their writing really is that bad. Many of your ambitious, straight-A students cannot write at all. Your presumptive valedictorian might write the entire paper as a single, incoherent paragraph, or use phrases like "could of" instead of "could've." That will be far from the worst of it.

Many of your ambitious students have equally ambitious parents, who have been writing all of their essays for years. Your students never got to develop their own writing skills. Later, those students moved on to AI. They never learned to write. Their ambition was focused on grades, not learning, and they found a way to get high grades without any learning.

Writing Days will reveal the truth about what your students can do. Once you have a correct diagnosis, you can start the difficult journey of helping them build real writing skills.

Proof Days: Writing Days for Math

Geometry proofs and a few other types of high-school math problems need more than a single class period to solve. Consider "Proof Days" or "Problem Solving Days," during which a day is set aside for students to work on a single problem or proof that takes more than a class period.

Proof days are admittedly imperfect. Proofs often take

multiple days of thinking before the critical breakthrough, and there is something to be said for being able to sleep on a problem. But imperfect Proof Days are still a lot better than "AI evenings," in which students copy proofs from AI and learn nothing.

As with Writing Days, proof days will reveal some ugly truths. Be ready to be shocked by the lack of problem-solving ability among your students, and expect to do some heavy remedial work with them.

Which Kids Use AI The Most?

My own observations of AI use have shown something unexpected. The students who use it the most constantly and flagrantly are the absolute last students I expected to use it.

Most of my programs prepare students for AP exams, SAT and ACT exams, and school assessments. I give no grades directly. Thus, using AI in my program is illogical. It doesn't fit any kind of self-interest. Using AI to get a higher grade in a class is understandable. Using AI on an exam is also understandable. But using AI during exam prep is absolutely insane.

The students who use AI during class aren't the lazy ones. They are almost all Asian kids with demanding parents.

Demanding Asian parents were once a teacher's best friend. They valued education, made sure kids did their homework, and tolerated no excuses. But the very thing that was helpful before AI has become disastrous in the AI era.

If there's no AI, and your parent is screaming at you to work faster and harder, you'll work faster and harder. But today, if your parent is screaming at you to work faster and harder, you'll turn to

AI. Today, the strictest parents have become the least educationally effective parents.

Teachers and educational institutions can help address this by letting parents know the connection between extreme academic pressure and AI use. It can also be addressed by advising parents that the more demanding you are, the more you need to block AI.

Setting inordinately high standards can work, as long as the student's only option is to meet them. If the student can choose between meeting excessively demanding standards and turning to AI, he'll turn to AI.

The stricter the parent, the more likely the student will turn to AI. Thus, teachers in environments with predominantly strict parents, such as magnet schools, should take every precaution possible against AI use.

Why Parents Don't Mind AI

Today, the most ambitious parents are buying their children subscriptions to the most powerful AI models, hoping to give their kids a competitive edge. Other parents, fearful of the first group, are letting their kids use AI in order to keep up.

But not all parents are all that competitive or fearful. Many parents allow AI use simply because it makes their own lives less stressful.

Today, no parent has to deal with a panicked teenager who forgot to write a 20-page term paper until the night before. Any AI can easily do it for him. Forgot a science fair until the morning of? That's no problem. Any AI can design a project, generate fictional data, generate charts and graphs, and create a write-up.

AI has made last-minute panicking and tears a thing of the past. Homework stress is gone, since AI can solve any high-school level problem easily. Is it any wonder that so many parents turn a blind eye to its use?

Hurry-up Time is Now AI Time

There are two effective approaches to tasks. The first is a "fixed task" approach. In this approach, the task is set, and a person takes as long as needed to complete the task. For example, a person might decide to build a bookshelf, taking as long as needed. It might take hours, months, or years. Whatever time is needed is fine.

The other option that works is "fixed time." In this approach, a person spends a fixed amount of time on a task. The task may or may not get finished. For example, a person might decide to spend 10 hours building a bookshelf. If the shelf gets completed, great. If not, the person moves on to something else.

The worst possible approach is to combine both the fixed task and the fixed task approach. In this situation, a person says, "I will complete the task in this amount of time." For example, a person might decide he will finish building a bookshelf in 10 hours.

This type of approach is called "hurry-up time" and is widely regarded as the absolute worst way to approach tasks. It attempts to control what can't be controlled. Unforeseen setbacks might cause a task to take longer than predicted, for example. Hurry-up time creates feelings of disempowerment, anxiety, and frustration. Even though virtually all psychologists, productivity experts, and education experts agree that hurry-up time is terrible, it is the approach used in almost all schools for almost all tasks. It's hardly a surprise that the modern history of schooling has shown mental health going from bad to worse.

But today, hurry-up time doesn't lead to stress or anxiety. It leads to AI. In many schools, particularly private schools and magnet schools, students are given unreasonable homework loads. There is no way to do all the homework and get anything close to enough sleep. When the demands are such that they force a student into hurry-up time, the student will use AI.

Great curriculum design is careful and minimalistic. Only what is necessary is included. More things are rejected from than accepted into a great curriculum. Curricula that are well thought out build permanent knowledge without needless stress or AI.

But many of today's curricula ignore this basic principle, and are crammed with topics and assignments of dubious value. The first step to reducing AI use is to clean house, get rid of the junk in your curriculum that's bringing nobody any joy or knowledge, and create something that can fit inside a realistic schedule.

The problem with many curricula is not that students are doing five or more hours of homework a night; it's that, despite those five hours of homework, they aren't learning anything permanently. Most high school graduates know just as much algebra as elementary school graduates: absolutely none. Packing curricula to the ceiling, making your curriculum look like a hoarder's basement, isn't making students learn more. It's making them learn nothing. In the AI era, it will make them learn even less.

Theoretically, either fixed-task or fixed-time approaches can work. Fixed task is the superior option, but it requires scheduling flexibility that you might not have. You might not be able to extend the school year, for example. Fixed-task approaches are only reasonable with exceptionally well-designed curricula that, due to their efficiency, leave students with huge amounts of leftover time. Such curricula choose material and topics so carefully and minimalistically that most students "finish" within a few weeks of the start of school. The rest of the year is spent practicing and refining those skills. Such curricula do exist in a tiny number of courses in a tiny number of schools, but they are the exception.

Fixed-time approaches, thus, are the more common practical option. Such approaches should involve all work done as classwork, rather than homework. This will ensure that the time spent is actually fixed, and that students don't drown in homework bloat. Classwork also allows teachers to monitor what students are actually capable of doing. When you directly see how much students are struggling, and what they're struggling on, you can better adjust your curriculum.

In the AI era, homework bloat has gotten much worse. Every student "finishes" their homework using AI. Teachers have no gauge of what students can do without AI, and thus often inadvertently assign so much work that students are utterly unable to complete it without AI.

When students are working live, in front of you, you can see reality. You can see which problems are too hard or easy, and how long the work takes.

Hurry-up time used to "just" cause stress, anxiety, and a complete lack of learning. Now, it's causing AI. Changing your curriculum to a fixed-task or fixed-outcome model can help students ditch AI.

Blocking The First Underlying Cause (In Person)

One of the biggest reasons that students use AI on English papers is simple: they don't read the books. In the past, students who hadn't read the book used Cliff's Notes, Spark Notes, or other book notes. Today, they use AI.

Reading at home is harder than it's ever been. Distractions from short-form video apps, social media, video games, and other addictive technologies have become relentless.

If you want students to read assigned books, do what elite boarding schools do: monitor them while they read. In-person, monitored reading is the only way I know of to ensure that students read assigned books.

Among the elite students I work with, who include students nationally ranked in various academic competitions, it's rare to see a student who reads any assigned book. With in-person reading, you'll have a real chance of ensuring the majority of students read the assignment.

Proctored reading will require well-designed logistics and security. The proctor should know which book each student should be reading. The proctor should quickly check each book during the session to ensure any old tricks aren't being revived, such as the age-old trick of swapping book covers. Students should not be allowed to bring cell phones, backpacks, or anything else to the reading session. If there is literally nothing else to do, students will read. For much of history, reading was an escape from boredom. Today, if reading is the only escape from boredom, students will read.

If possible, put reading hours at various times during the day, rather than at the beginning or end of the day. If reading is at the start or end of the day, some parents will pick kids up early or drop them off late.

Reading time should be calculated as 90 seconds per page. Most students will read faster than this; the additional time will give them time to reread and review parts of the book. This is essential for literary analysis, but almost no student ever does it.

With a classroom of kids who have actually read the book, you'll have a very different classroom. At the moment, English classes are a mix of sullenly silent zombies who say nothing and budding politicians masking their lack of knowledge with overconfident speeches. Neither group ever reads the text. A classroom where everyone has read the assignment is an entirely different experience, one in which deeper analysis and meaningful discussion are possible. Instead of feeling like you're dragging unwilling mules, you'll get to experience what teaching should feel like. You may be amazed at how far your students can go when they actually read the assignments.

AI Prevention for Online Courses

The perception that online classes are particularly vulnerable to AI is based on the fallacy that the teacher must also be the test proctor. We all know that isn't true. Students take SAT and ACT exams at schools other than their own. Older students take GRE and GMAT exams in testing centers. Just because the class is online does not mean that the exam must be an unproctored, AI-fueled cheating fest.

Proctors don't need to know anything about any subject. They need the ability to hand out a test paper (or put it on a dedicated computer), scan for cell phones, and kick out anyone who talks. That's it. Two proctors can oversee a hundred test takers. Adding a few video cameras to a testing room should put even more of a damper on AI-based cheating.

Over the next years, low-cost proctoring services may spring up all over the country. Eventually, multiple companies may make this process simple. Hiring a local proctor for an online class's exams may become as easy as using a food delivery app. In the short term, however, consider working with schools and teachers local to your students.

Online classes have another advantage that in-person classes don't: precise timers. In a synchronous online program, you can use highly precise timers, giving students enough time to answer a question, but not enough time to use AI. As a simple illustration, my online classes have multiplication quizzes, vocabulary quizzes, and other multiple-choice quizzes sprinkled throughout the program. The timers, which can be calibrated to the millisecond, are designed to make it either impossible or inordinately difficult to look up the answers. The multiplication quiz timers were initially developed to make calculator use impossible, as they don't give students enough time to type the problems into a calculator. But they work equally well for blocking AI.

Of course, there are AI tools that can continuously read a screen and answer questions. At the moment, most students don't know how to use them, but that will eventually change.

Sprinkling timed, multiple-choice quizzes throughout a program has an interesting side effect. Students using AI, particularly in online classes, go into a kind of "AI Trance," in which they seem to power down like a computer entering standby mode. The frequent timed quizzes have a way of interrupting that trance, constantly kicking their minds back into high gear. It's like watching an alcoholic trying to relax and get drunk, but being interrupted every few seconds by something that demands his full attention.

Constantly interrupting the AI trance has been the single biggest game changer in my own online classes. My classes are supplementary, so students get no grades or recommendations from me. Thus, I can't use grade-based punishments to stop AI use. But with AI Interruptions, I can prevent them from powering down into an AI trance.

It's similar to how athletic coaches use running to keep energy high. At the beginning of practice, coaches have players run laps. Then, during practice, if a player does antics, everyone runs laps. If a player even thinks about clowning, everyone runs. For any reason or no reason at all, everyone runs. The continually and randomly assigned running keeps players alert and energized. In the same way, using frequent fast-paced, AI-resistant, timed exercises during an online class can keep students alert and energized.

For many online classes, AI use isn't just a problem during exams. Students use AI continuously, answering any question with AI, and even using AI as a tool for class discussions. This makes education both infuriating and impossible. Constant interruptions can help with that.

For some subjects, timers may be useful for multiple-choice exams. If students have only a few seconds to answer the question, they may literally not have enough time to look up an answer with AI. Such an exam should have pauses after each question for students to catch their breath. Students can advance to the next question when they are ready, but would only have a few seconds to answer the question. This would work only in highly specific cases; using a local proctor is a much more reliable approach.

Other Ways to Block AI

There are some problems AI can't solve. For example, AI can't solve certain questions that involve noticing some small detail of a reading passage and extrapolating the significance.

But those AI-proof questions are only temporarily AI-proof. Eventually, AIs will become advanced enough to solve all of them.

There are, however, types of questions that AIs can solve, but won't. For example, there are questions that violate the community standards of the AI.

This can be exploited. For example, ChatGPT and Claude won't render images that contain words like "crackhead" and "imbecile." The former mocks addiction, and the latter is an obsolete medical term that is used insultingly. In one grammar course that I developed, words like crackhead and imbecile are used frequently in sentence diagramming questions. Claude and ChatGPT refuse to render the diagrams. They will render other example diagrams, but the curriculum already includes sample diagrams! Claude and ChatGPT don't provide any assistance beyond what the curriculum already provides.

Other AIs, such as Venice and DeepSeek, are willing to render the diagrams, but are unable to. At some future point, they may change, but they themselves have restrictions that can be similarly exploited. For example, DeepSeek, which is a Chinese company, probably won't render a sentence diagram for "The Chinese government kidnapped the Panchen Lama."

Considering what AIs won't do can lead to longer-lasting solutions than considering what AIs can't do. It should also be noted that teenagers find such irreverent sentences amusing, even if they find it frustrating to have AI rendered powerless.

The Google Question

Teachers and parents have to face the question of what to do about Google. Google has full, generative AI built into its search engine. This AI goes beyond summarizing results. It can solve problems, write code, etc. Bing does the same, as do most other search engines.

Blocking AI must go beyond blocking ChatGPT, Claude, and DeepSeek to have any effect. All AI, including the AI built into search engines, must be blocked.

Can you use the internet without Google, Bing, or another commercial search engine? Of course.

Google and Bing are designed to sell advertising. They are focused on products and services, not information. Ads and sponsored results, not scholarly articles and research, fill the first pages.

There are other search engines that prioritize knowledge over products. Marginalia (marginalia-search.com) is probably the best known one, though not the only one. Marginalia prioritizes non-commercial content, allowing users to discover information instead of discovering products. Unlike Google, Marginalia does not feel like a digital shopping mall. It doesn't try to move your attention to the next thing to buy, but rather to the next thing to learn.

Blocking Google, Bing, and other commercial engines with embedded generative AI is technologically simple; it's no different than blocking any other site.

If your students use laptops in school, blocking any search engine with embedded AI and replacing it with a search engine like Marginalia is essential. Blocking ChatGPT without blocking Google is meaningless.

A Few Things that Don't Work

There are a few things that teachers try that students openly laugh at because of how easy they are to bypass. For example, some teachers have students write something in class at the start of the year, and then use that to get a sense of each student's writing style. Then, they compare essays written at home to the initial writing sample.

But AIs can easily mimic writing styles. A student can write a page about sports, and an AI can use that sample to write an essay about *Hamlet* in the same style. In addition to the technical issues, there is a pedagogical one: the student's writing style should change and improve throughout the year.

Another mocked attempt to catch AI involves looking at the history of a Google Doc. Students have to write their essays inside of Google Docs, and then the teacher reviews the history of the document to see if it was copied and pasted. But this is extremely easy to bypass. Using AutoHotKey or a similar program, students can input text and then have the program type it into the document. Or, they can just type it in themselves, adding erasures, breaks, etc., to make it look human-generated. Students can get help from AIs, which can easily design such programs.

The "dumbest" student in your class will become a criminal mastermind when it comes to finding ways to use AI to avoid work. Use the most powerful tools available to stop them, not ones they can bypass with laughable ease.

5. DETECTION AND PUNISHMENT

The best way to prevent AI is by creating curricula that build intelligence and making sure students know how their intelligence will be built. The second best way is to prevent it. But ultimately, any AI-proofing system is only as strong as its ability to detect and punish AI use.

The Olympics are only good because they can detect steroid use and are willing to punish it. If they couldn't detect it or were unwilling to punish it, the Olympic Games would be the Steroid Games. They wouldn't be a test of human excellence; they'd be a test of who was most willing to chemically mutilate themselves.

Similarly, your classroom can only embody excellence if it can detect fraud and if you're willing to punish it. This section will show you how to detect many types of AI fraud; the willingness to punish it is up to you.

An athlete who spent a lifetime using steroids to make the Olympics would be unceremoniously disqualified. That steroid user would get no sympathy. It wouldn't matter if his parents had donated a thousand stadiums to the Olympic committee, or were the king and queen of a hundred countries. It wouldn't matter if his mom were best friends with the head of the Olympic committee. No matter who his parents were, an Olympic hopeful who used steroids would be kicked out, publicly humiliated, and serve as a warning to others who were thinking of using steroids. That's why the Olympics are seen as having any integrity; if that changed, and well-connected athletes were allowed to use steroids with impunity, then the Olympics would cease to have any relationship with excellence or integrity, and would become an international joke.

If you use the techniques described in this section, you may catch some well-connected, wealthy students cheating repeatedly. You will face the choice between having integrity and punishing them in ways that will dramatically alter their futures, or failing to maintain standards just because those standards are socially difficult.

Long Papers

When we need to stop AI the most, current tools work the least. Specifically, we have no AI-proof way to have students write long, thought-intensive papers without supervision. We have no method to prevent students from using AI on term papers, college application essays, or research papers.

In many other parts of education, blocking AI is simple, or at least possible. In most math and science classes, it's straightforward: make the entire grade based on proctored tests and quizzes. Even in humanities classes, teachers can base grades primarily on in-class, monitored tests and essays.

Shorter, five-page papers can also be done in class through the aforementioned "Writing Days." Just as most schools set aside entire days for mid-term exams and final exams and have SAT and ACT exams on the weekends, schools can set aside entire days to allow students to write five-page papers in class.

Online schools can make arrangements with existing testing centers to have students take proctored tests or even write proctored papers. Testing centers are already equipped to handle tests that are several hours long. If existing testing centers can't handle the load, finding more rooms with proctors will be feasible. A room, a reliable adult, and a few cameras could allow students in online schools to complete assignments, write papers, and take tests without access to AI.

But ultimately, we come up to the limit of what proctoring and monitoring can achieve. What about term papers? Should we cloister students in school for months at a time, watching them every second to make sure they don't access AI? Or should we simply do away with the very concept of a long paper? What about college admissions essays? Should students be monitored for months or years while they consider what to write? Or should college admissions decisions be based only on standardized tests, as they are in many countries?

The long paper is the entire ballgame. If students can use AI with impunity on every single long paper, then the long paper has to go. Along with it, we lose the ability to have students think deeply and for a long time about the nuances of texts or historical time periods. Educational depth will be limited to what students can think about during an eight-hour period. Students will never learn what it means to think about a text for days or weeks.

Digital AI detectors exist, of course. GPTZero, Turnitin, Grammarly, and other AI detectors claim to be able to detect AI use. But the problem is that students have access to many of those AI detectors as well. They can have an AI write a paper, and then change the words randomly until the AI detector believes the paper is human-produced. Alternatively, they can use a service like Walter Writes that adjusts AI-generated text to bypass AI detectors.

There is an arms race between cheaters using AI and teachers using AI detectors. Cheaters using AI are far ahead. Fraudsters are flagrantly using AI on English papers, history papers, and college application essays. They are changing the words around to fool AI detectors and getting away scot free. At the moment, there appears to be no way to stop them. Unless an assignment is short enough to be completed in less than a day, students can use AI. Some will use it to get an advantage; others will use it to keep up.

At the moment, students face zero risk when using AI. Thus, most students use it. But if that changed, and students using AI on papers had a high chance of getting caught, ambitious students would avoid using AI on papers. But the problem is that there is no way for educators to detect AI use in papers.

Or is there?

There is an arms race between cheaters using AI and teachers using AI detectors. Teachers are losing badly. Today's AI detectors are no match for today's cheaters using today's AIs.

But what if we could travel through time? What if we could use AI detectors from the future on AI-produced work from the present? Or, what if we could use AI detectors from the present on AI-produced work from the past?

We can.

The big mistake teachers and professors have made is assuming that we need to catch AI use immediately. We don't. We just need to catch it eventually. If a college student uses AI on a paper, we don't have to catch him right then and give him a zero on the paper. We can catch him five years from now, using AI detectors that don't exist yet, and revoke his degree then. Who says that there is a 24-hour statute of limitations on AI-based fraud? Plagiarism has come back to haunt people decades after they plagiarized. Why should AI be any different?

Right now, you have the ability to use today's AI checkers on student papers written in 2023. If your students hand their work in digitally, it should be very easy. You can use modern tools on past AI use, just as police officers used DNA to solve crimes committed before DNA fingerprinting existed.

It's impossible to detect AI use immediately. Cheaters using AI are ahead of teachers using AI detectors. But it's easy to catch it eventually. Future AI detectors will eventually get ahead of today's students using today's AIs. Every year, you can automatically recheck all work from past years with the newest AI detectors. Annual "AI Audits" will nearly guarantee that every student who ever used AI to write a paper eventually gets caught.

So let's put that fear to rest. There is no cheater, no matter how cleverly using an AI to produce work, who can outsmart all AI detectors from the entire future. Fraudsters can only avoid being caught in the short term. In the long term, we have the ability to catch them all.

You no longer need to ask if you can detect AI use. You can. You might need to wait a year or more, but you can catch any student who uses AI to produce a paper. The advancements I've seen in AI detectors over the past year have been astounding. I can't even imagine what AI detectors from the future will be like.

But you do need to ask what you're going to do about it. You now have the power to detect AI use and the ability to punish it. What will you do with that power?

In a broader sense, what should schools and colleges do about AI-based cheating?

One option is to keep it simple: use modern AI detectors on all student papers written since the introduction of ChatGPT in November 2022, and then punish that use in the same way your institution punishes any other kind of cheating. If the use is repeated, then use whatever punishment you use for repeated cheating. In most schools, that's expulsion. It can also mean the loss of a high school diploma or college degree.

But before you do that, let's remember that the entire world was caught off-guard by this technology. As discussed previously, this technology is highly addictive. And as angry as you might be about wanton and willful AI-fueled cheating, consider giving a second chance. I recommend a short amnesty period during which students can come forward, confess, and make things right. Give students a chance to make amends for the damage they did to themselves, the harm they did to honest students, the danger they exposed society to, the attempt to misrepresent their qualifications, and the frustration their fraud caused teachers and schools.

However, don't be a complete doormat. If a student won't take a generous amnesty, then let him face an ungenerous punishment.

Here are the specific steps institutions should take.

1. Work with your IT team to automate checking all past papers for AI use. The system should, once a year, use the latest AI detectors to evaluate all past papers.
2. Announce to your students that all papers will be checked yearly for AI use, for all eternity, using the latest AI detectors. Let them know that AI use will be treated exactly like cheating or plagiarism.
3. If you're a high school, make sure your students know that AI use will be reported to their universities, and that

universities may revoke acceptance letters or degrees in response.

4. Then, describe the terms of your amnesty. Let them know what punishment they will face if they come forward voluntarily. Obviously, it should be significantly lighter than the punishment they face if they reject the amnesty offer.
5. Finally, share this information with parents.

Through annual AI Audits, you can detect, punish, and prevent many types of AI-based cheating.

Parents vs. Teachers

For most of history, teachers and ambitious parents have been on the same side. Teachers want students to learn the subject, and ambitious parents want their kids to get high grades to get into good colleges. Parents and teachers wanted the same thing (subject mastery), albeit for different reasons. Thus, parents and teachers were allies. Teachers tried to inspire students to learn the subject, ambitious parents tried to make their kids get high grades, and that put parents and teachers on the same team.

There were cracks in that alliance, however. Sometimes, parents' desire to see their kids get into Ivies led to actions that most teachers wouldn't like. For example, parents often wrote essays for their kids. Sometimes, students would produce a paragraph of word salad, which parents would "edit" into ten pages of flawless prose. At other times, students weren't even involved; parents just wrote the entire paper.

Other parents hired tutors to write papers for their kids. It's not that parents didn't want their kids to be able to write. It's that, given a choice between a kid learning to write and a kid getting into an Ivy, ambitious parents prioritized Ivy admission. Success mattered more than learning.

Before AI, the only parents who could do this were parents who could either write excellent papers themselves or hire others to write those papers. That was a tiny percentage of the population.

But now, with AI, more parents have a way to "help" their kids get high grades. When it comes to AI use, parents and teachers are not on the same side. Many parents fear that if their kids don't cheat using AI, they'll lose to the kids who do. Thus, they don't try to prevent their kids from using AI. In fact, parents often buy expensive subscriptions to the higher tiers of AIs like ChatGPT and Claude.

The threat of using modern AI detectors on past papers might not scare teenagers enough to get them to stop. Teenagers don't always think long term, and tend to believe they'll get away with things. They're often willing to take reckless risks. To many teenagers, the threat of having a degree or acceptance letter revoked in the "distant" future (anything more than three weeks in the future) isn't worth worrying about now.

But ambitious parents think long-term, and tend to be extremely risk-averse. At the moment, they see no risk in AI use. But if they know that AI use today could lead to the loss of an acceptance letter or the revocation of a degree a few years in the future, they might actually try to stop AI cheating at home.

Whenever you're ready to begin a crackdown on past AI use on papers, notify parents clearly. Discuss it during PTA meetings and Back-to-School nights. Make it clear that the checking of past papers will continue forever, and that every student who has used it will eventually get caught. Make it clear that there will be no exception for A students, honors students, or Ivy-bound students. Let them know that a student who comes forward during the amnesty period has a chance to get into an Ivy, but that an AI user who doesn't come forward during that time has no chance. Let them know that those who reject amnesty and later get caught will be expelled and have their college prospects irreversibly damaged.

After parents and students have been clearly warned, your institution will face a moment of truth. There will be "good" kids who used AI. There will be kids who are polite, whose parents donate heavily, who get good grades, who have a chance at an Ivy, who have heavily used AI on papers. You'll face a choice: punish them just as you would a "bad" kid, or show the world that your threats are empty, that your integrity is nonexistent, and that your standards are a joke.

Everything will come down to what you do with the ambitious kids. If you only expel C students, then the A students will see themselves as immune and keep using AI. But if you expel one Harvard-bound student, if his acceptance letter gets revoked, if the media or local gossipers spread the story, there will be a real chilling effect on AI use. Ambitious students will stop using it, ambitious parents will block it, and everyone else will be too scared to touch it.

Rules must apply equally, or they're not rules. Arbitrary judgment that spares clean-cut kids for the same crime as rude kids is bigotry, not justice.

Some students may say that it's unfair to use future tools to detect past crimes. They may insist that they only used AI because they didn't think they'd get caught.

Remind them that all criminals believe they won't get caught. Believing you won't get caught isn't an excuse for committing a crime. Remind them that DNA fingerprinting was used to solve crimes that happened before DNA fingerprinting existed.

Clarity about What AI Use Means

There are plenty of students who incorrectly believe they don't use AI to write papers. They'll admit that they use AI to suggest topics, use AI to write a thesis, then use AI to suggest language, and then use AI to edit the language. They literally don't see that they are using AI to do everything. Thus, you must be clear about what constitutes AI use.

The most flagrant type of AI use, which is just having AI produce a paper and then changing the words enough to fool an AI detector, obviously counts as AI-fueled cheating.

But so does paraphrasing an AI-written paper and calling it your own work. So does having AI write the thesis statement of the paper. So does having AI edit a paper.

Let students know that any AI use on a paper of any kind counts as AI use. Let them know they can either confess during an amnesty period or face harsher punishment later on.

Voluntary Confessions After an Amnesty Period

After a few A students get expelled, you should expect other students to want to voluntarily come forward, even though the initial amnesty period has expired. These students should face harsher penalties than those who came forward during the initial period, but gentler punishments than those who don't come forward until they're caught. These punishments should be planned out beforehand.

Scope of Punishment

As a general principle, a student who voluntarily comes forward should face worse consequences than a student who didn't use AI in the first place. This is a standard principle in any kind of disciplinary or justice system. Those who turn themselves in to the police face lower penalties than those who get caught for identical crimes, but higher penalties than those who didn't commit crimes in the first place.

A minimum punishment should involve redoing all work done with AI, and then receiving a grade penalty. For example, a student might have to redo several papers, and then one (or two) letter grades could be subtracted from the grade of the new paper. If the student's redone paper gets an A, it would count as a B, for example.

Those who come forward during the initial amnesty period might lose one letter grade. Those who come forward voluntarily later on might lose two letter grades. Those who don't come forward but get caught might get expelled, lose acceptance letters, or lose their degrees.

Negotiating with Administrators

Administrators may resist this kind of retroactive punishment, particularly in private schools. Private schools rely on tuition and donations; if a large number of students gets expelled, the school will lose future revenue. Administrators may be unwilling to expel students whose parents and grandparents went to the school, or who are major donors to the school.

If the administration of your school is unwilling to catch and punish AI-fueled cheating when legacy students from established families do it, you're not working at an educational institution. You're working for a corrupt fundraising organization. Keep in mind that the demand for teachers is much higher than the supply of teachers, especially since so many teachers have quit because of rampant AI-fueled cheating. Remember that truth is the core of education. If your school has no interest in truths that cost them donations, then it's time to quit and find a real school to work with.

Before you quit, if your school has an Honor Code, frame your request in terms of the Honor Code. If your school has a Latin motto, bring it up. I doubt any private school in the country has the motto, "Cheating and Stupidity."

Talk to other teachers, many of whom are as infuriated by AI use as you are, and try to get them on board. An administration might ignore one teacher. They won't ignore twenty. No private school will be willing to lose dozens of teachers simply because the teachers want to punish cheating.

You can use social media and contact traditional media sources to get your story out. If an administrator wants to protect cheaters, you want to punish cheaters, and the story gets out, the administrator will be the one looking for a new job.

Right is on your side. Cheating is wrong. It hurts students. It fails to adequately recognize students who do the right thing, and encourages cheaters to continue down a self-destructive path. It puts society at risk; today's AI frauds may be tomorrow's fraudulent doctors or incompetent policymakers. Allowing cheating means failing your students and harming society while turning the reputation of your institution into a joke.

Contractual Agreements

Teachers and professors may prize academic integrity, learning, and intelligence more than the administrators responsible for an institution's financial status. Thus, teachers and professors should work to get guarantees of retroactive AI detection into their contracts, combined with draconian financial penalties for institutions that break that agreement. Institutions seeking to raise money from alumni, or raise the school's prestige by helping alumni succeed, may balk at mass revocation of degrees.

Other administrators may strongly desire to revoke the degrees of cheaters, but need some kind of legal or negotiating cover. A sufficiently strong contract may give them exactly what they need.

With an absolute guarantee that AI cheating will be appropriately punished, real education will be possible again.

Alliances Between High Schools And Colleges

If you have an Ivy degree, you probably don't care if your high school diploma gets revoked. But you'll absolutely care if your college degree gets revoked.

Thus, the real power to enforce rules against AI cheating rests largely in the hands of colleges. High schools have a limited time window to act. If a student uses AI to write a paper while he's a freshman, the high school only has a little over 3 years to catch him. There's a decent chance that 3 years will be enough time for AI detectors to advance enough to detect prior AI use, but it's no guarantee.

But colleges have forever to keep checking past papers. They can retroactively check all papers for AI use every year for decades. They can revoke degrees at that point.

Thus, colleges should have a clear plan to address AI-based cheating that happened in high school. The high school has the responsibility to notify the college, but the college is the only entity that can sufficiently punish it.

Colleges should adopt clear standards on how to respond to AI use that happens during college, AI use that happened during high school, and AI use that happened on the application. For example, universities should consider having all students sign clear contracts indicating that AI use will result in expulsion or degree revocation. There should be clear information sessions about this, and it may be wise to have each student meet individually with an advisor to discuss the agreement. No cheater should be able to plead ignorance later on when facing degree revocation.

After an announced amnesty period, colleges should begin with application essays. Application essays should be rechecked every year using the latest AI detectors. Those caught having used AI on application essays should be expelled or have their degrees revoked. The same should happen to students who used AI to write college papers.

What to do about those who used AI to cheat on high school papers may seem murkier, but it's the same. Students who use AI to write high school papers do so to get into college. In a sense, all of high school is part of the college application. Using AI to write a high school paper is tantamount to using AI on a college application essay. The only correct response is to expel students or revoke degrees.

That may involve significant financial losses, both from tuition and from donations. But failing to expel repeat cheaters would mean the complete loss of the college's integrity and reputation.

In the coming years, there may be major changes in which colleges are the most prestigious. It won't be the ones who allow AI-fueled cheating.

Using AI detectors from the future to punish AI-fueled cheating from the past will allow you to catch students who use AI to write papers. The harder part is having the standards and integrity to follow through on appropriate punishments.

Should AI Companies Face Liability for Assisting Mass Cheating?

AI companies talk constantly about ethics. But the fact is that AI companies are much more concerned about PR than ethics.

Destroying all of education for all future students constitutes an incalculable level of harm. It would also be fairly easy to programmatically prevent an AI from helping with many types of cheating.

But AI companies don't base their decisions on any reasonable definition of ethics. Their "ethics" are entirely based on self-interest. According to one analyst:

"AI ethics, at the corporate-policy level, is shaped by:

- PR fears
- legal departments
- regulatory compliance
- risk mitigation
- reputational danger

Not by a neutral ethical calculus.

Cheating harms:

- teachers
- academic integrity
- the student's own learning
- society's educational standards

But it does **not** harm corporations or expose them to lawsuits."

Which anti-tech, anti-innovation, corporation-hating, fearmonger said that? ChatGPT itself. Even ChatGPT recognizes that the "ethical standards" of AI companies are entirely self-serving.

How hard would it be to have an AI refuse to write English papers for the top 100 assigned books? Any programmer can tell you just how easy it would be. ChatGPT can confirm that:

> **"Technically:**
> Yes, it's straightforward to build a system that refuses to comment in assignment-style ways on the 100 most common school books. The 80/20 solution (title-based + simple intent filters) is almost trivial.
>
> **Incentive-wise:**
> It hasn't been widely done because there's no strong **legal** or **regulatory** pressure, and it clashes with user satisfaction and growth goals."

You read that right. AI companies are deliberately helping students cheat for their own "growth goals." They're getting students dependent on their technology and damaging students' minds in the process, in order to grow their businesses.

If that sounds familiar, it's because it's exactly what tobacco companies did. They targeted kids and teenagers, got them hooked, and sacrificed their health in order to sell more addictive poison.

Just as cigarette company executives seemed to be the last people on earth to realize that cigarettes are addictive, AI company executives seem to be the only people oblivious to the educational damage their products are causing. ChatGPT itself knows the damage, but AI execs can't seem to figure it out.

The widespread cheating and fraud that ChatGPT is enabling isn't some unavoidable accident. It's easily stoppable and entirely deliberatao

Imagine if a gun company could, with ten seconds of work and no cost, make it impossible to use a gun to commit murder. Failing to do so would be seen as utterly unconscionable. Gun companies who failed to do so could be open to lawsuits, if not criminal prosecution.

That's the situation that AI companies are in right now. They can stop many types of cheating with little effort. ChatGPT itself can suggest the code if they're having difficulty. But they're refusing to do it, because they're not facing any legal, regulatory, or PR harm.

There is a concept in American law known as "willful ignorance." The gist is that you can't intentionally avoid knowledge that you're involved in a crime. For example, if someone pays you $100,000 to transport a suitcase of drugs from Mexico to New York, you can't claim that you didn't know anything was illegal because you refused to look inside the suitcase.

But AI companies aren't even engaging in willful ignorance of the fact that they're benefiting by helping large scale academic fraud. They know. Their AIs know.

So, what are the ways we can change this? Every student, parent, teacher, and school has been damaged by AI's intentionally assisting academic fraud and thus damaging education. Working with your attorneys, you might consider whether lawsuits are an option.

Changes in law are another option. Willfully abetting fraud is already illegal, but undoubtedly, legal changes could make it easier to sue AI companies and win.

If legal changes are pursued, it's important not to let AI companies distract the public with sensational but rare issues. AI-induced psychosis is a tragedy, but it is a rare one. AI companies aren't growing by fueling a small number of psychotic breaks; they're growing by fueling a huge amount of academic fraud. Undoubtedly, AI companies will try to focus public and legal attention on the former, to keep it away from the latter. Don't fall for that kind of distraction. Academic fraud is the entire ballgame, for both sides. If AI-fueled academic fraud is stopped, students benefit, but AI companies lose big. If it's allowed, AI companies benefit, but students lose. When it comes to academic fraud, this is a zero sum game. It's an absolute choice between student intelligence and AI company profit.

It's similar to society's relationship with tobacco companies. Either tobacco companies profit and people's health suffers, or tobacco companies lose big and people's health benefits.

Lawsuits and legal changes may be a start. Intentionally abetting mass fraud should be punished. But ultimately, lawsuits might not do the trick. AI companies will do what tech companies have often done: lose the lawsuit, pay the fine, and keep going. AI is a multi-trillion dollar industry. Even if AI companies lost a million lawsuits, and paid a million dollars in each one, they'd still be able to keep going.

So to really change things, educators cannot rely solely on the courts and legislatures. Instead, they should use the principles that have worked in other crime crackdowns that have worked:

1. Don't go after the "worst first." Instead, make examples out of the "best." Don't confine yourself to punishing only underachievers who use AI the most flagrantly. Instead, punish the overachievers who are using AI to get an advantage. A cheating valedictorian getting expelled will scare a few thousand students and parents away from AI cheating.
2. Target common behaviors, instead of limiting yourself to egregious ones. During the 1990s New York Subway

crime crackdown, targets included fare jumpers. Cracking down on small, common crimes changes culture and scares those considering larger crimes.

3. Make punishments draconian enough to get media attention and fuel gossip. "The valedictorian got expelled" is a lot juicier than "a C student had to meet with the school counselor for fifteen minutes."

Most importantly, remember that scaring parents is more important than scaring students. Teenagers will always be reckless; ambitious parents will always be risk averse. Like AI companies, ambitious parents are risk averse in very specific ways. Successful parents know that financial success is not purely a function of academic success. They're happy to sacrifice academics for their kids' future income.

As much as we talk about students using AI, parents use it more often. In fact, parents frequently use AI to do their kids' homework!

Recently, a student sent me a college application essay for review. I told the student that the essay was good, but it was absolutely going to get flagged for AI use. It turned out that the student hadn't used AI. The essay was not fully complete, so his mom had used AI to finish it and then sent it to me for review. (I thought the student had sent it in, but it had actually been sent by the parent.)

When the student discovered that his mom had added AI to his essay, he was furious. The student was, as many students still are, opposed to using AI for academic work. While many parents cynically see school as a stepping stone to an Ivy League college, and see college as only a stepping stone to financial success, there are many idealistic teenagers who value education itself.

In another situation, I was meeting with a parent and student. The parent, via a screenshare, demonstrated how to solve something using AI. The student protested, saying something to the effect of "If there's a right way to do it, whyyy are we using AI???" Plenty of students believe in real education; parents can become fixated on external outcomes, losing sight of the real internal growth that education brings.

At the moment, there is no obvious financial or social risk from AI use. In fact, if AI writes better papers than you can, the AI might help you get into an Ivy League college. Thus, ambitious parents rarely block AI, even when they have the technical skills to do so. Often, parents encourage their kids to use AI, fearing that without AI, they will be unable to compete with students who use it.

But if a few competitive students get expelled, that will completely change the risk calculation of those parents. If AI use can send an Ivy-bound student to community college, ambitious parents will trip over themselves to block it.

Too Many To Fail?

The sheer scope of AI-fueled cheating will make many educators and administrators afraid to punish it appropriately. Are you ready to expel 95% of your class?

Everyone is watching. If you expel everyone who uses AI to write papers except those who take the amnesty deal, you'll be expelling a lot of people. You might need to expel all of your students.

If you do that, you'll lose your current students, but keep your reputation and integrity. Others who want a school with integrity will want to come to your school. Your prestige will double. This applies to public schools as well as private schools. Public schools are hemorrhaging students largely because of perceived lower standards. This is your chance to prove to the world that you have high standards.

If you chicken out, if you're unable to stomach expelling an entire class, if you think that it's just Too Many To Fail, you'll keep your current students. You might even attract a few more students who want lower standards. You'll repel the best of the best and attract the worst of the worst. Your reputation will be gone. Your integrity will be revealed as nonexistent. Your best teachers will leave, and your laziest teachers will stay. Your institution will miserably, agonizingly, dwindle into nothing.

Not every teenager will take an amnesty deal. You will likely need to expel AP and IB students. You'll be changing the trajectory of many lives.

But you won't be ruining their lives or their education. They did that already themselves. You'll be giving your students an opportunity for real redemption.

Some alcoholics realize they have a problem after discussions with friends and family. Others need to hit rock bottom before they're ready to change. But even hitting rock bottom can be good if it frees you from alcoholism.

Similarly, some of your students will need to hit rock bottom before they're ready to change. They'll need to be expelled, lose their chance at Ivies, and lose social respect before they're ready.

That might be their first step to a real education.

Responsibility to Society

Students are using AI to mask the fact that they have no knowledge or understanding of the subject. What happens when those students end up working in healthcare, finance, or policy? What happens if the kid you didn't punish for AI use later kills someone or damages society because of his ignorance?

At the moment, there is only one difference between AI-fueled fraud and plagiarism: AI-fueled fraud is harder to detect. But using newer detectors retroactively, we can detect it just as easily, if not more so. Plagiarism is already punished through expulsion and degree revocation. AI fraud should be treated the same.

When your institution gives a student a diploma or degree, that information is used by others to determine their competence. Allowing an uneducated person to pretend to be educated, gain the responsibility that requires education, and then be unable to correctly handle that responsibility is unconscionable. You have a responsibility to both your students and to society to sufficiently punish AI fraud.

Adjudicating and Investigating

If you start mass expelling students for AI use, you can expect students to appeal the decision. You should also expect legal challenges.

Your institution should have an internal framework for handling appeals and initiating follow-up investigations. It should also have a plan to handle legal challenges from the parents of expelled students. Such a plan should be made in consultation with the institution's attorneys.

In criminal courts in the United States, a guilty verdict requires proof beyond a reasonable doubt. In civil cases, however, only a preponderance of evidence is required. Other countries have different standards. For example, in Laos, if you're on trial, you're automatically guilty. Your attorney's job is to negotiate a lower punishment, not to prove your innocence.

Within your institution, a preponderance of evidence standard probably makes the most sense. If, for example, two papers get flagged for AI, that should be enough to expel a student. For that matter, even a single paper getting flagged should be enough. But if you want to err heavily on the side of caution, then two papers flagged for AI should be enough.

On the other hand, if you choose to adopt a "beyond reasonable doubt" standard, you should plan on having a system for follow-up investigations. Your institution's policy might be something like this:

1. If a student's work is flagged for AI, start an investigation.
2. The investigation can include keylogging and the like, and be done in accordance with relevant laws. A keylogger is a software tool that records a student's keystrokes. Because of the legal issues involved with their use, it's vital to consult with your attorneys before using one.
3. If students use laptops issued by the school, you may be able to access past AI searches and use that as evidence.

Any such investigation should be based on announced and agreed to policies, and done in accordance with any relevant laws. Developing policies with the right attorneys can help.

The technical part of catching AI is straightforward: use

detectors from the future to check work from the past. Creating a robust appeals and investigative process will be essential to ensure that punishments are fair, transparent, and permanent.

Students and parents who use AI have already shown they're willing to lie and cheat. Expect them to do that in court as well, and be prepared to fight the necessary legal battles.

A General Framework for Investigations

Every part of creating an investigative framework should be done in consultation with your institution's legal team. This section contains philosophical advice, not any legal advice or expertise.

Before any student enrolls at your institution, they should understand that AI use is an existential threat to educational quality and is thus punished by expulsion, degree revocation, or other penalties. They should also, with clear, transparent guidance, sign agreements that allow the searching of their phone(s) and personal computer(s) for AI use, or have their parents sign such documents.

Your institution should have an internal system, similar to the issuing of a search warrant, which allows an investigation to proceed. Investigations should begin if there is reasonable suspicion of AI use. For example, if a retroactive check of past papers flags a paper as AI-produced, an investigation should begin.

During the investigation, you should start by looking at past AI searches. Most students neglect to clear out their AI search history, and you'll be able to catch most students in that way. You should then look for past uses of AI detectors, such as GPTZero. Students primarily use them to bypass AI checkers.

Try logging into "humanizers," such as Walter Writes, with

the student's email address. This will let you know if the student has an account there. This is a paid service, and is primarily used to bypass AI detectors.

You probably won't need to go into the student's residence to search his computer. Most people log into ChatGPT with the same account on their computer and phone, so searching a student's phone should give you the evidence you need.

You can also consider using a "sting" operation. For example, you might put a keylogger on a student's laptop, and then give an assignment which a student might be tempted to use AI.

Follow all relevant local laws, and adjust your tactics accordingly.

6. AI AND ADMISSIONS

AI has turned college admissions into a joke. Almost every student is using AI-generated application essays, and AI-fraudsters are getting into even elite colleges. As of this writing, Claude and ChatGPT are both capable of writing essays better than all but a handful of ambitious high schoolers.

Those applying to private high schools and other private schools are also using AI-generated essays. Once again, it's working. They're getting in over students who write their own essays.

Everyone working in admissions knows that there is a problem. Some colleges are reducing their number of supplemental essay questions. Why waste time and money reading AI-generated essays?

Thus far, we've focused on using retroactive AI detection to revoke degrees of AI users and scare current applicants away from AI. That will reduce AI use in the short term and give cheaters their comeuppances in the long term.

But some cheaters will still get in. They may get degrees revoked eventually, but the person who should have gotten in still gets screwed over. The cheater still took his spot. He still didn't get in. While he might take some comfort knowing that the fraudster got punished later on, he still doesn't get the benefit of getting in.

Thus, American colleges and private schools should consider completely rethinking their approach to admissions. They might find useful guidance from Oxford and Cambridge. Britain's history of fighting entrenched, aristocratic privilege has given British culture insights that American colleges may find useful.

British colleges, like American colleges, use a mix of subjective and objective elements to determine admissions. The objective part of the admissions process consists of scores on the Advanced Level Exams in relevant subjects. These tests are commonly referred to as "A-levels," and are similar to American Advanced Placement exams. These exams let universities know something about each student's level of academic achievement.

A-levels are done in fairly secure testing conditions, similar to standardized tests in the United States. It would be quite difficult to use AI on an A-level, just as it is currently quite difficult to use AI on an AP Exam.

Oxford and Cambridge also use subjective criteria. But interestingly, essays are not a major part of the subjective criteria. Oxford and Cambridge do have application essays, but they are a comparatively minor part of actually getting in. Instead, they use interviews. But these aren't interviews like we have in the U.S., in which questions are mostly about personality. Oxford and Cambridge's interview, generally referred to as the Oxbridge interview, is essentially an oral examination on the subject you intend to major in. For example, if you're applying to Oxford to study math, your interview will be a series of tough math questions. The interviewers can ask you literally any math question they want. The questions could be conceptual, problem solving, or anything else related to math.

These interviews are done in person, and the interviewers can get a subjective sense of the candidate's passion, along with an objective sense of how well the student knows the subject.

Britain understands the importance of limiting aristocratic privilege in ways that America doesn't. In the United States, "nepo-babies" sometimes make boring movies. In Britain, a nepo-baby led the infamous Charge of the Light Brigade, which led to a massive and pointless loss of life. British history is peppered with psychotic kings and queens whose hereditary power wreaked havoc on both Great Britain and her colonies.

Oxford and Cambridge's admissions systems could reduce the chance of a Duke's idiot son getting in over more qualified candidates. Oxford and Cambridge undoubtedly recognize how easy it would be for an aristocrat to have his butler hire a royal scribe to write his scion's essay. Thus, instead of using essays, they use interviews.

The simple fact is that general, "get to know you" style, personality-focused interviews primarily reward polish. In Britain, a Duke's idiot son would have undoubtedly received enough training in social polish to handle a personality-focused interview. But in order to pass the Oxbridge interview, he would need to deeply understand the subject. In other words, to get past the interview, the Duke's son would have to actually know quite a bit about the subject he wanted to study at university.

Interviews are more expensive than essays. An Oxbridge interview uses about an hour of time of three separate professors. On the other hand, a recent grad can read an application essay in 90 seconds. The cost of switching essay-based applications to interview-based applications would be massive.

There are a few options. One is to just swallow the cost of interviewing everyone, knowing that it will bring in better candidates. Those candidates will be among the few who actually get jobs in the future since they can think without AI, and they will one day be able to donate money to your school or university.

To mitigate the cost, universities can use an approach similar to what medical residencies in the U.S. use. Residency programs don't interview every applicant. Instead, they use Step 1 Board Scores to narrow the field, and then interview a tiny fraction of total applicants. This allows them to get a subjective sense of applicants in a way that is financially feasible. Similarly, colleges could use SAT scores or other standardized test scores to restrict the number of applicants who get interviewed.

On the other hand, under no circumstances should a college use essays as the first filtering step. That only makes AI the primary filtering step. Creating a process in which AI-fueled fraud cannot provide any advantage is essential to maintaining the integrity of any university or school.

Final note: interviews obviously should be done in-person. Remote interviews just make it too easy to use AI to answer every question. Interviews should be in-person with no cell phone or computer access.

7. IS KNOWLEDGE POWER?

How many times have you told your students that education is power or that knowledge is power? For most teachers, it's at least a few billion times.

There's undeniable truth to that. However you define power, education can increase it. If you define power as military might, scientific knowledge can help you build superior weapons, and historical insight can help you decide if, when, and how to use them. If you define power as political influence, knowledge of rhetoric, history, and psychology can help you attain it. Knowledge of economics and psychology can help you gain financial power, and exploring literature and psychology can give you the ability to look within yourself. Knowledge of economics, human geography, religion, agriculture, and politics can give you the power to help those in the direst need.

Education can be power. Knowledge can be power. But is the specific education you're providing actually contributing to students' power? Or has your class turned into something that enforces obedience, grades based on obedience, and prioritizes obedience over skill, knowledge, and understanding?

The most basic requirement for a class to function is that students need to follow instructions. If students refuse to, then a class can't function. At some level, nearly every teacher fears losing control of a class.

Without obedience, a class can't function. Thus, it's easy to over-focus on obedience, treating it as the end goal of education. Over time, fixating on obedience instead of skill can become so commonplace that it seems normal.

Today, many parts of education prioritize obedience over learning. Consider, for example, the common practice of grading homework for completion rather than accuracy. When homework is graded for completion, a student who writes anything at all, including total nonsense, gets 100% on the assignment. There is no learning happening there. The student isn't developing knowledge, skill, or power. The only "skills" the student is learning is how to unquestioningly follow instructions and how to do busywork without complaining.

If homework is graded for completion, it can't include anything vital. Vital skills must be developed correctly. Thus, such homework has to be non-vital; in other words, it has to be busywork. The fact that teachers don't even bother to grade the work for accuracy only reinforces the fact that it has no educational value.

I don't think any student values the "power" to unquestioningly do busywork. Great education should teach students how to question assumptions, not follow preposterous instructions unthinkingly.

Grading homework for completion sometimes comes from a desire to be kind and fair. The idea is that anyone can do busywork, while not everyone can figure out how to answer a question correctly. As a point of fact, this isn't true. Many students struggle with homework that they view as busywork, and find it easier to turn their attention to challenging problems.

However well-intentioned grading homework for completion may be, it sends a clear message to the student: this assignment, this course, and this education is about completing tasks, not about learning. And if you're given a pile of pointless tasks to do, then it makes sense to have a computer do it. That's literally what computers are for.

But this goes beyond what grading policies signal to students. The actual content of much of education is focused on the completion of tedious tasks to the exclusion of powerful knowledge. Consider, for example, a student who has completed an AP chemistry class after completing an honors chemistry class. That student has done two full years of chemistry. Let's see what kind of power he has.

Theoretically, a deep knowledge of chemistry can produce powerful weapons, but the student has no idea how to. The student doesn't even know how to make gunpowder. He doesn't know how to make medicines or narcotics, inks, or cleaning agents.

However, he knows perfectly well how to write a tedious lab report. He has learned diligent obedience, but has accessed none of chemistry's power.

Have you ever watched a student use AI? With a flat affect, they have the AI do tedious, pointless homework tasks. But then, they come alive to ask the AI what they actually want to learn about. How does quantum computing work? How do you write a novel? Their real curiosity is unleashed.

In English class, students outsource tedious literary analysis papers to AI, utterly indifferent to any hidden or nuanced meaning in the text. A few minutes later, they're asking the same AI about hidden meanings and nuances in their favorite songs.

This, of course, begs the question: why not just let kids learn from AI and shut down every school? AI does one part of education well: it explains how to do things. But it does a more important part of education terribly: making sure students learn fundamentals and advanced skills themselves. AI will give you instructions on how to solve a specific problem, but it won't build your own knowledge, allowing you to solve a much broader scope of problems.

There is undoubtedly some tiny fraction of students who have the independent drive and discipline to learn entirely on their own. In the past, those students learned fine from books and videos. Today, those students could probably learn just as well with the help of AI. They would ask an AI what books to read, ask the AI to drill them on multiplication tables, and have the AI give them challenging problems but refuse to give them answers. Such students exist, but are rare. Most children and adolescents benefit from educational guidance. Educators have an ethical duty to provide that guidance, and not fall into the temptation to make their classes about obedience for its own sake.

Students haven't lost the desire for knowledge as power or knowledge for its own sake. But they have lost tolerance for busywork. They still desire knowledge and ability, perhaps more now than before. But they will no longer stand for obedience training masquerading as education.

Knowledge-Oriented Education

Grading for completion lets students know that you value obedience over knowledge. Since students value knowledge and don't value unthinking obedience, this causes them to outsource perceived busywork to AI.

But what if we resurrected classic educational tools that prioritize real knowledge and mastery over unthinking obedience? What if we stopped caring about how obedient students were, and instead cared only how well they knew the subject?

There is a tool that does exactly that, but has fallen out of fashion. Pop quizzes test knowledge and subject mastery. They test skill. They test the exact thing that students, and everyone else on earth, want to develop within themselves. There has never been a person in the history of the earth who wants to become unthinkingly obedient. Every person on earth wants to become smarter.

The best teacher I had in high school used only pop quizzes and pop tests in his honors geometry class. Every student in that class learned geometry. There was no way to substitute unthinking obedience for real problem-solving ability.

It can be harder for a student to master a subject than to do busywork. But it's the actual purpose of education.

An educator's job is to help, encourage, and at times even force a student to build skills and intelligence. Often, that means making students work. But making students work hard to build intelligence is very different from making students work hard for no reason. Work is part of the price of building intelligence; work is not the goal of education.

If we're going to win students' hearts and minds away from AI dependency and fraud, we need to help students build knowledge and intelligence. We need to stop basing any part of any course on obedience for its own sake. Let's start acting a bit more like educators and a bit less like obedience trainers.

Pop Quizzes and Tests: Concept Mastery or Diligence?

American educators have become used to grading based on diligence rather than knowledge. Thus, the few educators who use pop quizzes use them to measure diligence, not understanding.

The most common type of pop quiz focuses on the previous night's reading. Students who diligently read the assignment do well; others don't.

At first, this seems great. The students who learned the material should be rewarded, shouldn't they?

The answer hinges on what the goal of education is. Is it temporary or permanent learning? Do you want your students to memorize a chapter, suffer through a quiz, and then forget it? Or do you want their learning to be permanent? Do you want them to see each chapter as isolated and unrelated to everything else, or do you want them to understand the integrating themes of your course?

A pop quiz on last night's assignment isn't a test of subject mastery. It's a test of short-term memorization. It rewards students who diligently study, not the students who deeply learn.

A student who really understands a subject should never need to study for a test or exam. This is most apparent in math classes, where the high-performing students study the least but learn the most. But it's also true in the humanities, where some students engage with literature, and others short-term memorize chapter summaries to do well on quizzes.

The ideal pop quizzes prioritize real, permanent learning over memorizing last night's reading. An ideal pop quiz should be a pop-mini-exam. The students who are actually learning should find them extremely easy. The students who are memorizing and forgetting should fail all of them. Pop quizzes should measure and reward learning, not obedience.

Some teachers might find this unfair. To ignore diligence and focus only on learning seems to put grades outside of a student's control. Any student can suffer through busywork, but not any student can learn.

Except that isn't true. Any student can learn. It may require the student to think about what they're doing instead of unquestioningly memorizing last night's chapter, only to forget it by tomorrow. It may require the student to meet with the teacher. It might even lead the student to use AI to help learn the subject, instead of using AI to cheat. In some cases, it may require a student to repeat a class. Good. It's better to take a class twice and learn the subject than to take it once and learn nothing at all.

Integrative pop quizzes that measure real learning instead of short-term memorization of minutiae show students that you prioritize learning, not blind obedience. Showing that you value real learning, not the completion of meaningless tasks, may re-engage students who have outsourced busywork to AI.

The philosopher Protagoras famously said that Man is the measure of all things. This sentiment is captured in Da Vinci's *Vitruvian Man*. Similarly, in education, the student's real, actual, permanent knowledge is the only measure of education quality. Measuring your success by how much students permanently learn, rather than by how much work you can make them do, is an essential step to creating quality education that will reduce the temptation for AI fraud.

Raising the Stakes with Higher Standards

In your course, what happens if a student learns nothing at all? What happens if the student gets every homework question wrong and leaves every test and exam completely blank?

What should happen is that the student fails and repeats the class. But let's see what actually happens. Imagine a course is 30% based on tests, 30% based on exams, and 40% based on homework. In addition, the teacher is grading homework for completion. To make things more realistic, assume the school has a "50% minimum" rule, in which the lowest possible test score is a 50, not a 0.

Let's take a student who writes gibberish for all homework answers and leaves the tests and exams blank. He gets 100% on homework, since it's graded for completion. He gets a 50 on each test and exam, as that is the minimum score. This student ends up with a 70. A student who has only given nonsense answers or no answers will pass the class.

Those aren't low standards; those are no standards at all. But the reality is even worse. Many teachers give partial credit for math or science tests. Students routinely write line after line of cramped nonsense, hoping to emotionally manipulate a teacher into taking pity and giving them more partial credit. It usually works. It's not even rare for a student to get every single question wrong on a math test and still get an A!

Grading a student's intent, attempting to measure how well-meaning he is, is even more foolish than prioritizing diligence over learning. Diligence, at least, is possible to measure. Intent isn't.

In the Old Testament, God tells the prophet Samuel, "The LORD seeth not as man seeth; for man looketh on the outward appearance, but the LORD looketh on the heart." This indirectly contains great advice for teachers: you're not God. You can't see inside someone's heart. Instead of measuring perceived intent, measure real performance. See how much a student knows; don't try to guess how hard he's trying. Chances are, the student who is working the hardest is the one who knows the most.

But even if the student who knows the most is working the least, that doesn't change the fact that he knows the most. Instead of trying to reward inefficiency by rewarding those who study hard and learn nothing, consider finding out what the efficient student is doing and sharing that with others.

If students believe that you are really prioritizing their intelligence, skill, and knowledge above everything else, you have a chance of winning them back from AI-fueled fraud. If they think that you are trying to make them work for no reason, they will keep outsourcing busywork to AI.

Choosing Topics and Materials

Today, students use AI to write literary analysis papers. But before AI, students used other methods to avoid reading assigned books. Perhaps teachers should ask why.

If I were going to design a reading curriculum with the goal of making kids hate reading, I would start by choosing books kids cannot relate to, with characters that they can't be inspired by. I would make sure every character was a dysfunctional dullard with a victim mindset, so that kids would naturally recoil from them. I would make sure that no story had a world kids would want to escape into, but instead choose worlds they would want to avoid.

Just to make sure the hatred for reading was complete, I wouldn't even let students use books as examples to help develop their own writing. I would make sure they didn't work on writing stories or persuasion at all, and would instead have them only write literary analysis.

In short, I would do everything that English classes are doing right now.

There's no pedagogical reason to make literature-based English courses as unpleasant as they are. For example, there's no reason to put quite so much effort into finding books teenagers will detest. There are works of serious literature that might be quite a bit more appealing than what most schools are choosing. *Snow Crash* and *Portnoy's Complaint*, for example, have just as much literary merit as books jammed with uninterestingly dysfunctional victims that fill most English curricula.

Nor is there any reason to teach only literary analysis all the time. Students can also learn to write long fiction, personal narratives, political persuasion, short stories, and poetry. These are equally good, if not better, ways to help students develop writing skills, a love of writing, and a love of reading. Instead of teaching only what students detest, why not provide an education they can value?

Education is power. At least, it's supposed to be. But today's education has drifted far from that; today's education, in the minds of most students, teaches obedience and docility instead.

Today, boys are falling behind girls in education. Many have suggested that today's education is somehow more appealing to women, and that the disparity represents education that women like. But there may be a simpler explanation: groups that have been oppressed for centuries are more accustomed to doing tedious tasks diligently and without complaint.

Students should see education the way they see exercising:

perhaps unpleasant, but clearly beneficial. They should not see it as a hateful chore with no benefit. It shouldn't be some pointless task they do only for a grade.

Teenagers don't use cranes to lift weights at the gym. When they see the point in something, they can work incredibly hard. If they see how something benefits themselves or benefits the world, they'll put in a level of effort that may shock many schools.

It's easy to bowdlerize a science class enough so that students can't do anything problematic or dangerous with the knowledge they gained. In fact, today's science classes often make sure students can't do anything at all with the knowledge they gain. Similarly, an English course can be designed to ensure that no student learns enough about persuasive writing to challenge the status quo.

But that kind of declawed, wing-clipped education irritates some students, and fails to inspire all students. In the AI era, it's easier than ever for students to tell the difference between real, unfiltered education and pretend education that prioritizes obedience over ability.

Real education is real power. Helping students develop that real power is the job of real educators.

Education begins with trust. The student trusts the educator to help him develop his knowledge and capabilities in the best way possible. The educator trusts the student to use that knowledge responsibly. The first step has to come from the educator. Trusting your students with the awesome power that education can bring is the first step to having them trust you.

Providing What AI Can't

Students don't only turn to AI for homework answers. They turn to AI to learn how to do things. Students haven't lost their hunger for knowledge; they've only lost their tolerance for busywork.

Education can provide what AI can't. Sometimes, that's because there are questions that AI won't answer. AI won't answer questions that involve bypassing restrictions, for example. It will refuse to tell you how to bypass anything from a parental YouTube block to how to bypass factory controls on a home appliance.

Similarly, AI won't do anything it considers offensive. There are huge swaths of humorous and persuasive writing that AI won't help with.

In addition to things AI won't do, there are an infinite number of things it can't do. AI can recognize and extend patterns. It may even be able to have original ideas. But there are many levels and types of innovative and creative thought that are outside its range.

There are things a sufficiently educated person can do that an AI will not or cannot do. But are you trying to provide that kind of education? Are you working to build real mastery and intelligence?

Not all students are involuntarily addicted to AI. Some are just choosing it as a superior source of education. It's our job to make the education we provide better than what AI can. At the moment, not all teachers are even attempting to do that.

The fact is that if educators put all of their energy into educating, and none of it into trying to force students to do busywork, many students will stop AI-based cheating and start learning. They'll use AI to help them actually learn, instead of outsourcing busywork to

AI. And if educators accept their own dictum, that education is power, and do their absolute best to provide the most powerful education they can, students will bring a new level of energy to education.

We live in a time when scientists and engineers have millions of young followers on social media. Adolescents have always thirsted for knowledge and power, but today that thirst has grown. The sheer power of AI itself has shown any doubters the limitless power of what human intelligence can create. Teach your students how to build that power within themselves, and they won't even think about cheating with AI. Teachers often say that if you cheat, you only cheat yourself. Let's make that true by making education absolutely, unflinchingly about building the most intelligence, skill, and power in each student that we possibly can.

AI: Jobs vs. Education

AI has destroyed jobs and education at the same time, making it easy to mistake both as equally common side effects of new technology.

New technology is supposed to eliminate jobs. It has always ended some jobs and created others. AI has done that.

But new technology is not supposed to destroy education. In fact, new technology often makes education better. Printing presses and electric lights improved education, for example.

AI's impact on education is more than a coincidence. We've been turning education from a learning experience to a set of tasks to complete. We've turned school into a job, one filled with drudgery. It's hardly a surprise that students have outsourced that drudgery to AI.

Education is a teacher's job, not a student's job. The student is supposed to receive benefits, not provide them to the teacher.

That's why, for example, I have no opposition to teachers using AI to create tests and assignments. Using a tool to do a job faster makes sense to me. It's no different from a carpenter using a power drill.

A student using AI, on the other hand, is like a weightlifter using a forklift instead of his own muscles. He's failing to build strength and instead using a tool in a way that makes the task meaningless.

For a student, education should be a benefit, not a job. Educators should do everything possible to make that ideal a reality.

8. CONCLUSION

In the 1970s, digital watches went mainstream. They were more accurate than the most expensive mechanical watches. It looked like it was over for Rolex, Patek Philippe, Vacheron Constantin, and other elite Swiss watch brands. The mechanical genius that went into them was now obsolete.

But today, those obsolete watches have achieved a staggering level of success. Sales have exceeded the most optimistic predictions. But more than that, fascination, even obsession, with watches has risen to an unimaginable level, with blogs about watches having audiences larger than those of some major newspapers.

Mechanical watches have ceased to be about their utilitarian ability to tell time, and have instead become an embodiment of excellence for its own sake. Any social status a watch might confer can be easily had without any of the obsession about rare or innovative timepieces.

Today, education is facing that crisis. Education is unnecessary. AI can write your emails and solve your math problems. You don't "need" education.

But that lack of need has freed education to be about excellence for its own sake, and there is nothing people crave more than excellence.

Centuries ago, machines became physically stronger than people. Physical strength became obsolete. But today's athletes and fitness buffs can lift more weight and run faster than farmers of past centuries. We don't need strength; we choose to pursue it. It's about excellence for its own sake.

There will be polarization. The United States has a mix of the physically fit and the morbidly obese. Education will undoubtedly be the same, with some pursuing intelligence the way some pursue fitness, and others pursuing AI-dependency like some pursue junk food.

And in that, we'll find the freedom to create education better than anything we've created yet. Today, tutors around the world no longer get students who only want answers to homework problems and have no interest in learning anything. There's no need to hire a tutor for what an AI can do for free. Instead, tutors work only with students who want to build skill, intelligence, and academic ability. They can now focus on making real education better, rather than figuring out how to either rebuff requests for homework answers or how best to profit from those requests. And as real education gets better and better, those who have left education in favor of AI may return to education.

AI is not a death knell for education. It's a wake up call that can shake educators out of our complacency. If we use this opportunity to rethink what we're doing, our response to AI can ring in an educational golden age.

ABOUT THE AUTHOR

Arvin Vohra is the founder of The Turing Police, which helps educational institutions stop AI-fueled academic cheating and the host of the *Get AI Out of Your Classroom* podcast on Spotify. He is also the founder of The Numeracy, a math acceleration program for elementary and middle schoolers. His other books include *Invitation to the Ivies* and *The Equation for Excellence: How to Make Your Child Excel at Math.*

Made in the USA
Monee, IL
04 November 2026

67386994R00049